ANIMAL GENIUSES

DOLPHINS

THERESA EMMINIZER

PowerKiDS press.

Published in 2026 by The Rosen Publishing Group, Inc.
2544 Clinton Street, Buffalo, NY 14224

First Edition

Editor: Theresa Emminizer
Book Design: Michael Flynn

Photo Credits: Cover, p. 1 RuqayaMai/Shutterstock.com; (series background) Viktoriia LiSa/Shutterstock.com; p. 5 toshiharu_arakawa/Shutterstock.com; p. 7 F Photography R/Shutterstock.com; p. 9 Wonderful Nature/Shutterstock.com; p. 11 Nicolas-SB/Shutterstock.com; p. 13 KateChris/Shutterstock.com; p. 15 Yann hubert/Shutterstock.com; p. 17 Ewa Studio/Shutterstock.com; p. 19 JorgeSanz/Pacific Press/Alamy Live News; p. 21 ForGaby/Shutterstock.com.

Cataloging-in-Publication Data

Names: Emminizer, Theresa.
Title: Dolphins / Theresa Emminizer.
Description: Buffalo, NY : PowerKids Press, 2026. | Series: Animal geniuses | Includes glossary and index.
Identifiers: ISBN 9781499451856 (pbk.) | ISBN 9781499451863 (library bound) | ISBN 9781499451870 (ebook)
Subjects: LCSH: Dolphins–Juvenile literature. | Dolphins–Behavior–Juvenile literature.| Animal intelligence–Juvenile literature. | Dolphins–Psychology–Juvenile literature.
Classification: LCC QL737.C432 E45 2026 | DDC 599.53–dc23

Manufactured in China

CONTENTS

Delightful Dolphins 4
Part of the Pod 6
Teamwork 8
Clever Communication 10
Big Brains 12
Using Tools. 14
Making Medicine 16
The Mirror Test 18
Be Kind to Dolphins 20
Glossary 22
For More Information 23
Index 24

Delightful Dolphins

Dolphins love to play and have fun together as they swim in the deep blue sea. But these animals are more than just cute. They're supersmart, too! Dolphins are widely believed to be some of the smartest animals on Earth. What makes them so special? Read on to find out!

Part of the Pod

Dolphins are social animals, meaning they like to spend time with each other! A group of dolphins is called a pod. There may be as many as 30 dolphins in a pod. Members of a pod take care of one another. They work as a team to **solve** problems.

Teamwork

When a dolphin is sick or hurt, the pod will push it to the surface (top) of the water so that it can breathe. Pods also work together to attack, or fight, animals that **threaten** them, such as sharks. Dolphin pods have been seen swimming in **formation**, like dancers!

Clever Communication

Dolphins **communicate** with one another in many ways. They make noises, such as squeaks and grunts to share their feelings. Each dolphin has its own unique, or one-of-a-kind, **whistle**. Dolphins make a clicking sound when using echolocation, a way of gathering knowledge about what's nearby by sending sound waves into the water.

Big Brains

Like people, dolphins have big brains. Dolphins' brains are special in another way too. While dolphins sleep, one half of their brain stays awake while the other half shuts down! This allows dolphins to stay **alert** and rise to the surface to breathe while asleep.

Using Tools

Dolphins work together to catch food. They herd fish together or push fish onto shore, making them easier to catch. They also use tools! Some dolphins use a kind of soft animal called a sea sponge to cover their nose like a glove. This keeps them safe from scratches while they dig in the sand for food!

Making Medicine

Dolphins also use certain things as **medicine**! Bottlenose dolphins rub against corals and sponges to keep their skin healthy. The corals may help heal cuts on their skin and scrape, or rub, off harmful things they've picked up while swimming. Older dolphins teach their young which things are best to use.

The Mirror Test

Mirror tests are a way for people to learn about how intelligent, or smart, animals are. Very few species, or kinds, of animals can **recognize** themselves in mirrors! Most think their own **reflection** is another being. Dolphins recognize their reflection, meaning they are self-aware, or know who they are.

Be Kind to Dolphins

Because they're so smart and playful, dolphins are often trained to do tricks and perform in shows. But some people believe that keeping dolphins in **captivity** is cruel, or hurtful. You can be a friend to dolphins by doing your part to keep oceans clean and safe.

GLOSSARY

alert: Watchful.

captivity: For an animal, the state of living somewhere controlled by humans—such as in a zoo or an aquarium—instead of in the wild.

communicate: To share ideas and feelings through sounds and motions.

formation: A certain shape or pattern.

medicine: Something taken to fight illness.

recognize: To see and know something or someone.

reflection: An image seen in a mirror or shiny surface.

solve: To find an answer to a problem.

threaten: Likely to harm.

whistle: A high-pitched sound.

FOR MORE INFORMATION

BOOKS

Riggs, Kate. *Dolphins*. Mankato, MN: Creative Education and Creative Paperbacks, 2025.

Sexton, Colleen A. *Dolphins*. Minnetonka, MN: Kaleidoscope Publishing, Inc., 2023.

WEBSITES

National Geographic Kids: Bottlenose Dolphin
kids.nationalgeographic.com/animals/mammals/facts/bottlenose-dolphin
Read more fun facts about dolphins!

San Diego Zoo: Dolphin
animals.sandiegozoo.org/animals/dolphin
Find out more about dolphins.

Publisher's note to educators and parents: Our editors have carefully reviewed these websites to ensure that they are suitable for students. Many websites change frequently, however, and we cannot guarantee that a site's future contents will continue to meet our high standards of quality and educational value. Be advised that students should be closely supervised whenever they access the internet.

INDEX

B
brain, 12
breathe, 8, 12

C
captivity, 20
communication, 10

D
dolphin whistle, 10

E
ecolocation, 10

F
food, 14

M
medicine, 16
mirror test, 18

P
pod, 6, 8

S
sharks, 8

T
tools, 14
tricks, 20